Esteban Favela-Chávez

MANAGEMENT, PRODUCTION AND CHARACTERIZATION OF GRAPE VARIETIES

Esteban Favela-Chávez

MANAGEMENT, PRODUCTION AND CHARACTERIZATION OF GRAPE VARIETIES

FOR RED WINE

ScienciaScripts

Imprint
Any brand names and product names mentioned in this book are subject to trademark, brand or patent protection and are trademarks or registered trademarks of their respective holders. The use of brand names, product names, common names, trade names, product descriptions etc. even without a particular marking in this work is in no way to be construed to mean that such names may be regarded as unrestricted in respect of trademark and brand protection legislation and could thus be used by anyone.

Cover image: www.ingimage.com

This book is a translation from the original published under ISBN 978-620-3-03551-3.

Publisher:
Sciencia Scripts
is a trademark of
International Book Market Service Ltd., member of OmniScriptum Publishing Group
17 Meldrum Street, Beau Bassin 71504, Mauritius
Printed at: see last page
ISBN: 978-620-3-51334-9

Favela-Chávez, E. 2021. Handling, production and characterization of grapes for red wine. UAAAN UL, Torreón Coahuila, Mexico, p.

Content

I.- Introduction .

Grapevine is an economically important fruit crop worldwide, with *Vitis vinifera L.* being the species that dominates commercial grape production. In addition to this species, it is known that there are about 60 other species in the genus *Vitis*, distributed mainly in the northern hemisphere.

It is the oldest species in the world and is an ancient plant that produces grapes. Most of the grapes used, either as table fruit or for making wine or raisins, are of the *Vitis vinifera* species, said to be native to the regions between the southern Caspian and Black Seas in Asia Minor. Thousands of vine varieties have been derived from this species, and wine has accompanied human beings since ancient civilizations.

Wine is the beverage resulting from the alcoholic fermentation of grape must. There are factors such as the winemaking process, the variety and the method of cultivation, which mark differences in the attributes of a wine; however, the so-called terroir, influenced by luminosity, altitude, latitude, rainfall, slope, orientation and soil type, is what defines the typicity of the grape and the wine, which is expressed in the sugar content, acidity, color and aroma, among others. In general, the quality of a vintage is determined by the interaction of the cultivar with the soil and the prevailing climate in a region in a given year.

Mexico is considered the oldest wine producer in Latin America. Its wine industry provides close to seven thousand direct and indirect jobs and generates a turnover of just over US$40 million. Likewise, grape production in Mexico is aimed at the table, raisins, winemaking, and the production of concentrated juice. Wine production is one of the main activities of viticulture, the different varieties producing red wines are of good quality, descendant of *Vitis vinifera L.*, unfortunately it is very sensitive to phylloxera, so its exploitation must always be on resistant rootstocks. On the other hand, it is considered that wine consumption in Mexico is 535,000 hl per year, of which more than 70% is imported and per capita wine consumption is only 530 ml, national wines are produced in different geographical areas with different climatic and geographical characteristics.

On the other hand, wine is considered to be a beverage containing more than a thousand substances, most of which (such as vitamins and minerals) come from grapes. The beverage is obtained from the alcoholic fermentation of grape juice, which is produced by the action of yeasts that transform the sugars of the fruit into ethyl alcohol and carbon dioxide. Wine is known only as the liquid resulting from the total or partial alcoholic fermentation of grape juice, without the addition of any substance.

II. Literature Review .

2.1. Origin and history of the vine .

Weaver, (1976), mentions that the grapevine (*Vitis vinifera L.*) is the oldest species in the world and is an ancient plant that produces grapes and is frequently mentioned in the bible. Most of the grapes used, either as table fruit or for making wine or raisins, are of this species, *Vitis vinifera*, said to be native to the regions lying between the southern Caspian and Black Seas in Asia Minor, which has been carried from region to region by civilized man to all temperate climates and more recently has been cultivated in subtropical climates. Thousands of vine varieties have been derived from that species. Vinifera is also a progenitor of many hybrid vines obtained in the eastern United States.

Winkler, (1980) and Ferraro, (1984), consider that nowadays, the vine is cultivated in warm regions all over the world, being the largest producers: Australia, South Africa, the countries of Europe (Italy, France, Spain, Portugal, Turkey and Greece,) and in Mexico and Argentina. The lines of expansion of wine varieties were different from those of table grape and raisin varieties, due to the differences in customs and religion between the peoples of the southern and northern coasts of the Mediterranean. The vines spread to the Far East via Persia and India. Many years later, when Europeans colonized new lands, the vine was always among the plants that accompanied them.

Weaver, (1985) mentions that *Vitis vinifera* was brought to Mexico by the Spanish and to areas now occupied by California and Arizona. The vines introduced by the missionaries prospered and some of them grew to a good size. The English colonists brought vines from the Old World making plantations along the Atlantic coast in the colonies of Massachusetts, New York, Pennsylvania, Virginia, North and South Carolina and Georgia. With the beginning of Mexico's independence, the dependence of the crop began as a consequence of the prevailing political conditions and struggles, in spite of the attempts of Cura Hidalgo from his curato de Dolores, who was determined that vine cultivation should flourish in that land. Humboldt affirmed that in Dolores and San Luis de la Paz "there were vineyards, which Father Hidalgo certainly wanted to maintain. The struggles that exhausted Mexico for long decades frustrated any possibility of viticulture flourishing to the extent that in the region of Dolores vineyards almost completely disappeared, although in the northern regions of Parras, Coahuila, to whose initiation and development Lorenzo García contributed in the San Lorenzo hacienda in 1597 and which gave rise to the founding of the Villa de Parras, they remained in a state of survival thanks to the fact that vineyards and wineries acquired great importance as suppliers to the surrounding cities.

Villanueva (2011) mentions that from colonial times until the early 1960s, vine cultivation was located in four main areas of the country: Aguascalientes, the Lagunera region, Parras de la Fuente, Coahuila and Baja California Norte (BCN).

However, beginning in the 1960s, viticulture began to gain new spaces in the central and northern regions of the country and new states appeared that would eventually become the bastions of Mexican viticulture, such as Sonora and Zacatecas, whose contribution to the unusual dynamism experienced by viticulture in the 1960s was very significant.

2.2. World grape production.

Bravo, (2010) and Pérez, (2015), mention that grape production is an agricultural activity that has been carried out for a long time almost everywhere in the world, with European countries as traditional producers and exporters; however, in recent years there has been an increase in the production of table grapes in countries such as Chile, which is the second largest exporter of this fruit in the world. There is a clear ranking of the countries with the highest exports in the world: in first place is Italy, in second place is Chile and in third place is California. According to FAO figures, it reached 67.7 million tons in 2008, with a growth of 11.2% in the decade 1999-2008, although it remained quite stagnant in the last five years of the decade considered. The OIV also records a similar world production figure for 2008 and also establishes a wide variation in the geographical share of production over the last two decades. Europe, the world's largest producer, has lost a significant percentage of its share in world production, falling from 63.3% to 44% in the period, a share that has been captured by the rest of the world. Asia shows great progress in its share, almost doubling it from 13.9% to 26.5%. The Americas, on the other hand, recorded an increase from 17.3% to 20.7%, as did Africa, which increased its share from 4% to 6%, and Oceania, from 1.5% to 2.8%.

 Likewise Pérez, (2015), consider that this is an agricultural activity that has been carried out for a long time in almost all the world, having as traditional producers and exporters the European countries; however in recent years an increase in the production of table grapes has been observed in countries such as Chile, which is the second largest exporter of this fruit in the world. Exports during the same year totaled 1,735,414 tons. Of these, 610,000 tons were exported by Italian producers, 490,000 tons came from Chile, 215,000 tons from the United States, while Greece and Spain shared a figure of 100,000 tons exported. Turkey, which is one of the world's largest producers of table grapes, exported only 28,000 tons.

2.3. Area under vine in Mexico .

The National Association of Winegrowers (2008) mentions that the origin of wine is something very old that is confused with the origins of human civilization and is closely linked to the way of living and thinking of the people who harvest it and the region where it is produced.

Today, there are approximately 41,000 hectares in Mexico dedicated to growing grapes for wine production, with those in Baja California, Zacatecas, Coahuila, and Querétaro standing out, producing approximately 27,000 tons of grapes in each agricultural cycle, with a total of 23 wineries. Mexico is the oldest wine producing country in Latin America, and its main competitors in this continent are the United States, Chile and Argentina. The climatic factor is the main problem or obstacle to wine production in Mexico. Our country has approximately 42,000 hectares established with vineyards. This area is distributed in 14 states with the following percentage participation: Sonora 47%, Baja California 13%, Zacatecas 12%, Comarca Lagunera 10%, Aguascalientes 7% and Querétaro 4%, in addition to small areas in the states of Chihuahua, Guanajuato, Puebla, Oaxaca, Hidalgo, San Luis Potosí. However, the history of grapevines and wine has a long history linked to the first discoveries made by man about the chemical reactions of fermentation and oxidation.

On the other hand, El Siglo (2012) describes the current situation of wine, where it mentions that it has remained practically unchanged in recent years. Wine production in Mexico continues to be lower than the amount of imported wine. Approximately 65% of the wine consumed in Mexico comes from abroad. In this sense, we can consider the wine market in Mexico as a totally growing market, where every year both domestic wine production and foreign wine imports increase, as well as wine consumption. On the other hand, it should be noted that Mexico currently has a significant number of brands (more than 2,000 labels, 1,200 of which are Spanish) for the level of consumption in the country. This fact causes a situation of market saturation. According to data from the Servicio de Información Agroalimentaria y Pesquera (SIAP) portal of the Secretaría de

Agricultura, during 2007 Mexico imported 83,182 tons of fresh grapes and up to October 2008 recorded entries of 53,611 tons (Asociación Nacional de Vitivinicultores, 2008). In 2010 the planted area was 28,209 hectares, of which 67.2% is in the State of Sonora, 14% in the State of Baja California and 12.7% in Zacatecas.

Robles et al, (2004) mention that Sonora produces around eighty percent of the grapes in Mexico; in particular, 74% of table grapes, 98% of raisins, and 74% of industrial grapes. Thus, of the total hectares harvested in the State, 47% corresponds to table grapes, 35% to industrial grapes and 18% to raisins, where, in spite of being areas too hot for wine growing, the temperature is still appropriate for grape growing, mainly due to the altitude at which the areas of Querétaro, Aguascalientes, Sonora and Zacatecas are located. In these regions, the climate is Mediterranean, with wet winters, dry and temperate summers, with non-extreme temperatures and ideal characteristics of sun and rain. Grapevines do not thrive in tropical climates, but it should be noted that the exact demarcation of these bands does not necessarily exclude other regions, as is the case with Mexico in particular. Imports are mainly made in the last months of the year, especially from October onwards, and the main sellers of the product are the United States with 58% of sales to Mexico and chile with 42%.

2.4. Wine grape production in Mexico.

Tarango, (2015), refers that Mexico is the third largest country in Latin America in terms of territorial extension, after Brazil and Argentina, and the second in terms of population, which amounts to 104 million inhabitants. The rural population is estimated at 21 million people, 60% of whom live in extreme poverty. Of its total surface area of approximately 2 million km2, only a little more than 10% can be used for agricultural production. Its arid and semi-arid desert zones cover approximately half of its territory. Arid zones are considered to be those areas that receive an average annual rainfall of less than 350 mm, and semi-arid zones are those that receive between 350 and 600 mm per year. In both cases, the average annual precipitation is less than the maximum annual potential evaporation, thus evidencing a water deficit. These regions are characterized by a scarcity of water, with a highly erratic distribution of rainfall, occurring in few events and of a torrential type. This situation substantially limits the development of agricultural activities.

2.4.1. Northwestern Region.

It is mentioned that in Baja California, an area known as the wine belt, its climate favors harvests due to its humid winters and dry, temperate summers. Sonora: is the state where most of the land is devoted to grape growing. Despite having a hot desert climate and low rainfall, thanks to the use of irrigation systems, most wine production is concentrated in the State of Sonora.

Robles et al, (2004)), mentions that since 1999 Sonora has participated with more than 70% of the harvested area and in the last 10 years has produced an average of 72% of Mexican grapes, which places it at the head of this activity. In the last five years, the grapes produced in Sonora have gone from being destined primarily for industrial use (1995) to being consumed as fruit. In 2007, 88% of Sonora's grape production was destined for fruit, compared to only 9% for industrial use.

Likewise, Armenta Cejudo (2004) mentions that the State of Baja California generates more than 90% of the country's table wine production. Sonora has the disadvantage of producing raw materials for about eighty percent of the grapes in Mexico; in particular, table grapes produce 74%, raisins 98%, while industrial grapes produce 74%. Of the total hectares harvested in the state, 47% corresponds to table grapes, 35% to industrial grapes and 18% to raisins. Likewise, in the last five years, production has represented an average of 10% of the national total, although it began to decrease in 1998. Production in this state has traditionally been for industrial use (81% in 2007), in which the wine industry stands out, and as of 2000 the percentage destined for consumption as fruit began to grow.

On the other hand, Garritz (2011) mentions that in Sonora, table grapes are produced mainly in the municipalities of Hermosillo and Caborca, the former being the most important. In total, for 2008 the area produced 15,089,697 cases of grapes in all its varieties while the Caborca region produced 4,568,720 The Napa Valley in California is one of the most prestigious; it is 8 km at its base and

56 km high and has three different microclimates. The south side has the coldest climate. The middle region of the valley is a bit warmer and Cabernet Sauvignon, White Riesling and Chenin Blanc grow well there, while Zinfandel and Petit Sirah are planted in the warmer, northern end. This is where the Beringer company is producing, explains Kent Kantz, director of the company's laboratory in St. Helena, California.

The Embassy in the (2012), refers that wine production in Mexico in 2010 was 38,000 tons, according to the Food and Agriculture Organization (FAO) Despite the fact that Mexico currently represents less than 1% of world wine production, more and more, local and international producers are investing in Mexican wine regions, since according to some Mexican experts in the sector, wine consumption in Mexico is increasing by 13% annually. However, as stated in the first part of the study, consumption is still much lower than that of beer or other alcoholic beverages. The grape varieties produced in Mexico are Pinot Noir, Cabernet Sauvignon, Merlot, Garnacha and Alicante, among others. On the other hand, the Mexican wine industry currently offers 200 different types of wines, produced in the 18 states that are dedicated to grape production.

2.4.2. Northeastern Region .

López, (1987), and Tournier, (1911), mention that this area is one of the oldest and most recognized as a producer of quality table wines. The main grape varieties found in these vineyards are Cabernet Sauvignon, Merlot, Siras, Tempranillo,

Sauvignon Blanc, Semillon, etc. Phylloxera has been reported in this region since 1889, so the use of rootstocks is mandatory. This region has been characterized by the quality of the wines produced in it, being Siras, a variety that has adapted very well to the climate and soil conditions.

The Asociación Nacional de Vitivinicultores A. C. (2008), considers that the conditions in the Parras region are very special. (2008), considers that the conditions in the Parras Region are very special, in spite of being a semi-desert climate, the proximity of the Sierra Madre Oriental and an altitude of 1500 meters above sea level, causing warm days and nights, which translates from the viticulture point of view into ideal conditions for the production of high quality wines. In the Comarca Lagunera, viticulture began in 1925 and from 1945 it acquired regional importance, so that from 1958 to 1962 the vineyard area increased significantly. Coahuila: extremely hot climate during the long summer with abrupt temperature changes during the short winter season.

2.4.3. Central region.

Meraz, (2013), mentions that according to some sources and testimonies, it is indicated that the appearance of the first vines in the American continent arose 500 years ago, before the landing of Christopher Columbus in that territory, where in the year 1000, in an area of St. Lawrence, now Canada, led by Leif Eriksson, they established a colony called Vinland, which means "Land of the vines". However, the introduction of the vine or also known as vitis vinifera in America was brought by the Spanish conquistadors from the island of Cuba from Spain, where at the end of the conquest of Mexico they found wild vines, such as vitis lambrusca, vitis rupestris and vitis berlondieri. Mexico was the first territory where the first vines from Spain were planted. In March 1524, Hernán Cortes, after the conquest of the Aztec empire, ordered each colonist to plant one thousand Spanish and native vines for every one hundred Indians in his service, achieving an expansion of vineyards in Puebla, Michoacán, Guanajuato, Querétaro, Oaxaca, San Luis Potosí, Sonora and Baja California. Zacatecas, due to its climatic conditions, produces fine grape varieties rich in sugar and fast ripening.

On average, it contributes 9% of Mexican grape production, although in 2003 it had an atypical participation of close to 2% of national production. Unlike Sonora, in Zacatecas grape production is mainly destined for industrial use, which in 2007 accounted for 65% of production compared to only 35% destined for consumption as fruit. The varieties produced are Cardinal and Red Globo.

De la Cruz, *et al.* , (2015), mention the State of Querétaro, Mexico, located in the central zone of the country, at least two wine regions are distinguished, one of them corresponding to the Valley of San Juan del Río, whose climate is semi-dry temperate and where flat, deep soils with sandy-clay tendency predominate; and the other region is located within the municipalities of Tequisquiapan and Ezequiel Montes, where there are greater aridity conditions, with soils with slopes and more calcareous.

In Zacatecas, table grapes are produced in the municipalities of Ojo Caliente, Fresnillo, Luis Moya, Guadalupe, General Pánfilo Natera, Villa Hidalgo, Cuauhtémoc, Loreto, Calera, Villa González Ortega where the most important for table grapes is Fresnillo. Aguascalientes: the harvest areas are located in a wide valley between two mountain ranges, with a temperate climate with summer rains and a soil with a large amount of soluble salts.

De la Cruz *et al*, (2015), refer to the fact that after suffering a severe crisis in the 80s and 90s due to high production costs, low grape prices, lack of knowledge of production technology and imports, the Queretaro wine industry has reemerged, with a tendency to increase the established surface area and to improve the quality of its grapes and wines, mainly reds, for which new plantations have been established with fine varieties such as 'Merlot', 'Cabernet Sauvignon', 'Syrah', 'Tempranillo' Querétaro: area of fertile land with optimal climatic characteristics for vines, located at 2.It is located at 2,000 meters above sea level and has extreme conditions that range from 25° C during the day to 0° C at night.

2.4.4. Climate .

Joachim, (2004), considers that many arid regions are located under areas of high pressure, where rain-bearing frontal systems can rarely penetrate, consequently, these areas have scarce and scattered precipitation patterns, with large annual and seasonal variations. Rainfall occurs infrequently in the form of isolated thunderstorms, which can then produce flooding in dry river systems. The variability of these events often leads to periods of several years of drought or above-average rainfall. There is a large daily and seasonal variation in temperature. Cloudless skies and dry air allow the terrain and lower atmosphere to become very hot during the day. When this heat radiates back into the atmosphere after sunset, there is a sharp nighttime cooling and even freezing in winter. Air masses are usually stable and wind speeds are often low. Surface heating at some points, combined with the absence of trees and open landscapes, can produce strong local winds and high-speed eddies or dust devils.

2.4.5. Soils .

Joachim, (2004), refers that soils in arid and semi-arid zones can be shallow or deep, sandy or clayey, and can vary in acidity and fertility. Productivity depends on the soil's ability to retain water, which tends to increase with thickness and organic content. Sandy soils have a lower water retention capacity than clay soils. The sparse vegetation that is prevalent in arid and semi-arid areas allows water to dislodge soil particles into the soil pore spaces, which causes the soil to harden and absorb less water (soil exfoliation). This leads to more runoff and erosion of small particles, which contain nutrients, from the superficiency. Once eroded, the soil is less able to support vegetation and more susceptible to further modification by water and wind. The geomorphology of many arid regions creates large inland drainage basins with no natural discharge. Evaporation of water leaves salts in soils. If precipitation does not dissolve these salts and redistribute them, salinization of the land will occur.

2.5. Vine morphology .

The vine (*Vitis vinifera L.*) is a plant belonging to the Ampelidae family, described by Monlau as a family of sarmentose and climbing shrubs with stipulate leaves, opposite at the bottom and alternate at the top (Hidalgo, 2006). It has a group of vegetative organs such as roots, trunk, shoots, leaves and a group of reproductive organs, flowers and fruits. In the case of the former, its main function is to maintain the life of the plant by absorbing water and minerals from the soil, this to manufacture carbohydrates and other nutrients in the leaves, also influences respiration, translocation, growth and other vegetative functions. In the flowers, these in turn produce seeds and fruits. It is composed of two individuals, one constitutes the root system (*Vitis spp.* of the American group, mostly), called rootstock or rootstock and the other the aerial part (*Vitis vinifera L.*), called scion or variety. The latter constitutes the trunk, the arms and the shoots that carry the leaves, the clusters and the buds. The union between the two areas is made through the grafting point. The whole is what is known as the stock.

2.5.1. Root.

Reynier, (2001), mentions that the root is the subway part of the plant, it ensures the anchorage of the plant to the soil and its nourishment in water and mineral elements. Throughout its development, the root branches to form a network of roots called root system. The roots of the vinifera species are sensitive to phylloxera, so it is necessary to graft it on resistant rootstocks. The extent of the root system depends on the species, planting frame, soil type and cultivation techniques. Ninety percent of the root system develops above the first meter of soil, the vast majority being between 40 and 60 cm deep. There are two types of roots: the main root: present in plants grown from seed, which is pivotal, and the adventitious roots: in plants grown from cuttings, they grow laterally on a stem, the appearance of adventitious roots is usually in the subway part.

2.5.2. Stems and branches .

Ticó, (1972), mentions that the stem can reach considerable dimensions, is always wavy or twisted and is covered by an accumulation of old bark from successive years. Each year the overwintering buds of the vine develop, giving rise to a herbaceous shoot called shoot buds, which is a branch with variable long internodes, simple leaves arranged in an alternating-distic position, with buds in their axils. Opposite these in the third or fourth node is the inflorescence, in *Vitis Vinifera* they appear opposite two consecutive leaves.

Reynier (2001), on the other hand, mentions that a vine plant or vine, vine stock or vine, shows that it can have very varied forms and that the stems of an abandoned vine crawl along the ground until they find a support to which they can cling. The vine is, in fact, a vine, because it is necessary to regulate the regulation by a severe pruning and to impale it if one wants to raise it above the ground. The vine is therefore quite clearly distinguishable from other fruit species.

These parts are generally considered to be constituted by *Vitis Vinifera*, the stem of a cultivated vine (or plant) comprises a trunk, some main branches or arms and some herbaceous shoots or branches, if it is in a period of vegetative activity or some significant shoots which are the shoots (production) if it is in periods of rest. The trunk can be more or less defined depending on the training system. The height depends on the training pruning, being normally between 0.0 m in a Manchego vase and 2.0 m, in the case of a vineyard. The diameter can vary between 0.10 and 0.30 m. It is twisted, sinuous and cracked, covered externally by a bark that peels off in longitudinal strips.

2.5.3. Sheet .

Ticó, (1972), indicates that they develop in all the nodes of the shoot, in the axil appear the buds of the shoot. They have a lobed shape and five nervures and as many lobes; they start from the insertion of the petiole with the blade and constitute the support of the green part of the leaf, called blade. The leaves are inserted in the nodes are generally simple, alternate, distichous with an angle of 180 ° and normal divergence, composed of petiole and blade, petiole inserted in the stem, sheathed or widened at the base, with two stipules that fall prematurely. The coloration varies from light green to very dark. The surface is sometimes smooth, rough, wavy or curled, this surface, on the underside, is covered in varying amounts by hairs or fuzz, giving it a woolly or whitish appearance.

Leaf blade: generally pentalobulated (five nerves that start from the petiole and branch), with more or less marked lobes depending on the variety. With serrated edge; more intense green color on the upper side than on the underside, which also has a more intense hairiness, although there are also glabrous leaves. The size of the leaves is very variable in the same vine, and the general shape varies from one type of vine to another, which is taken into account for the classification, there are rounded, kidney-shaped, heart-shaped, wedge-shaped or truncated.

2.5.4. Stems.

Hidalgo, (2002), mentions that the tendrils are structures comparable to stems, they can be bifurcated, trifurcated or polyfurcated, with a mechanical function and with the particularity that only the tendrils that are rolled up are lignified and remain. They have a supporting or climbing function. The origin of the tendrils is the same as that of the inflorescences, and it can be considered a sterile inflorescence. The tendrils occupy the same position as the inflorescences, at a node of the branch and on the opposite side of the leaf, and quite often have several flower buds, which are rolled up. They have a clinging or climbing function. The tendrils and inflorescences have a similar origin, so it is common to find intermediate stages. In fertile vine shoots, the tendrils are always located above the clusters. The most frequent distribution of tendrils and/or inflorescences in the vine is the regular discontinuous one. The end of the free tendrils curl upwards forming a kind of spiral on itself, but when they find a support on the side in front of it, they curl upwards, as a consequence of the uneven growth of its parts. Meanwhile, the tendril that does not curl remains green, but when it does so, it intensely lignifies, giving support to the vine.

2.5.5. Yolks .

It is mentioned that the buds are inserted in the node, above the axil of insertion of the petiole. There are two buds per node: the normal, thicker bud that generally develops in the cycle following its formation, and the early bud that can sprout the year of its formation, giving grandchildren of less development and fertility than the normal shoots. If the early bud does not sprout during the year of its formation, falls with the first cold, it does not survive the winter period. All vine buds are mixed and axillary. The normal bud is more or less conical in shape and consists of a main vegetative cone and one or two secondary vegetative cones. These cones are formed by an embryonic stem, in which the nodes and internodes are differentiated, the foliar outlines and in its case, the outlines of the inflorescences, and a meristem or cauline apex at its end. These vegetative cones are protected internally by a cottony fringe and externally by two scales.

On the other hand, Reynier (2001) mentions that in the vine shoot, the bud is a shoot embryo, which is constituted by a vegetative cone ending in a meristem and provided with leaf outlines. Several types of buds are observed on the growing green shoot: At the extremity, the terminal bud, which ensures the growth in length of the shoot by cell multiplication and differentiation of new internodes, nodes, leaves, buds and tendrils; it falls at the growth stop. At the level of each node and in the leaf axil, a bud is able to develop rapidly shortly after its formation on the shoot, and a dormant bud is found on the shoot in winter. A bud in its development originates a shoot, which in autumn takes the name of the dormant buds have an essential function in the perenniality of the plant that allow the vine to develop new shoots each year. Each bud contains the outlines of the first organs that appear in spring and a terminal meristem that ensures the growth of the shoot and the new formation of new axillary buds.

2.5.6. Flowers.

Morales, (1995) and Ticó, (1972), refer to the fact that the flowers of *V. vinifera* are hermaphrodite, grouped in clusters. They have 5 sepals, 5 petals, 5 stamens and an ovary with two cavities each containing two ovules, the flowers are self-pollinated, there are sterile and fertile flowers depending on the species. If during the flowering period the temperature is low, the sun is insufficient, the soil is too wet and there is a lack of nutrients, the exchange of pollen can be obstructed and cause flower drop. The temperature required for flowering is variable and most occupy more than 20 ° C. It is a cluster inflorescence, initiated in late spring and summer in the year preceding flowering and fruiting. The main axis of the cluster is called rachis, and the individual flowers have a pedicel, a calyx with five sepals, a corolla with five petals, five stamens and a pistil with a short style and an ovary with two loculi. The inflorescence of the vine is known by the name of raceme, it is a compound raceme (raceme of cymes). The raceme is an oppositifolio organ, that is to say, it is located opposite the leaf. The cultivated vine bears from one to three clusters per fertile shoot. Normally there are two clusters and rarely four. Vines cultivated for their fruit are usually hermaphrodite. The flower is inconspicuous, small, about 2 mm long and green in color. The flower is pentamerous, formed by: Calyx, consisting of five sepals welded that give it a dome shape. Corolla, formed by five petals welded at the apex, which protects the androecium and gynoecium detaching at flowering. It is called cap or calyptra. Androcecium, five stamens opposite the petals consisting of a filament and two lobes (thecae) with longitudinal and introrse dehiscence. The pollen sacs are located inside. Gynoecium, ovary succumbent, bicarpelate (welded carpels) with two ovules per carpel. Style short and stigma slightly expanded and depressed in the center.

2.5.7. Fruits.

Morales, (1995), emphasizes that the fruit is a fleshy, succulent berry of variable flavor, color and shape. According to the variety, it contains from one to four seeds, although there are seedless varieties. The skin is covered with a layer of waxy cells called pruina that protects the fruit from insect damage, water loss and gives good appearance, the skin contains most of the constituents of color, aroma and flavor of the grapes and is richer in vitamin C. The seed is rich in oils and tannin and is the means of sexual propagation, although it is generally useted only for genetic improvement. The fruit has different shapes: spherical, ellipsoidal, obovate, ellipsoidal elongated, ovoid or oval. The bunches have different shapes depending on the variety and we can find: short conical, conical with shoulders, long conical, cylindrical, cylindrical with wings, conical with two wings of variable size more or less spherical or oval, and on average 12 to 18 mm in diameter. Red grapes contain a high percentage of tannin, a chemical substance that confers a bitter taste to the tissues in which it is found. This can apparently impair the quality of the fruit, as in the case of the Criollo grape, but it is an important substance for winemaking, it favors its fermentation and gives it what is called body.

2.5.8. Grape seed.

Hidalgo, (2002) and Calderón, (1998), refer that the seeds are the seeds surrounded by a thin layer (endocarp) that protects them. They are rich in oils and tannins. They are present in number from 0 to 4 seeds per berry. The seedless berry is called apirena berry. Externally there are three different areas: beak, belly and back. Inside we find the albumen and embryo. Once fertilization is completed, the result is the grape seed or berry, which fattens rapidly and is made up of an outer film, the skin; a pulp that fills almost the entire berry; the seeds and the prolongation of the channels of the short stalk, called the brush, which is why the flow of sap that feeds them all is carried out. Until well advanced in vegetation, the grain is green, has chlorophyll, that is to say, it produces at least part of the sap that nourishes it.

2.5.9. Pulp.

Togores, (2006), mentions that it is the most voluminous part of the grape berry, representing 75 to 85% of its weight, being formed by a typical vegetal parenchymatous tissue whose origin is the walls of the ovary, with large cells occupied almost all its volume by vacuoles, where the must accumulates, it represents the largest part of the fruit. The pulp is translucent, except for the red varieties (which accumulate their coloring matter here), and is very rich in water, sugars, acids (mainly malic and tartaric), aromas, etc. It is covered by a fine network of conductive bundles, being called brush to the extension of the pedicel bundles.

2.5.10. Husk.

The skin is considered to be the outer part of the grape berry. Its mission is to enclose the plant tissues that contain the reserve substances accumulated by the fruit during ripening, as well as to protect the seeds as perpetuating elements of the species until they reach their full development and to defend these structures from external aggression. The skin is formed by 6 to 10 layers of cells, is the outermost part of the grape and as such, serves to protect the fruit, which is membranous and has a cutinized, elastic epidermis. On the outside there is a waxy layer called pruin. The pruin is responsible for fixing the yeasts that ferment the must and also acts as a protective layer.

2.5.11. Pips or seeds.

Togores, (2006) and Castrejón, (1975), consider that the seeds constitute the element in charge of perpetuating the individual by sexual way, it comes from the ovules of the flower after fertilization. The external shape of the seeds allows to distinguish an almost flat dorsal face with two pits separated by the raphe, and a ventral face with a bulging groove and the chalaza, both terminated by the beak.the seeds are the seeds surrounded by a thin layer (endocarp) that protects them. They are rich in oils and tannins. They are present in number from 0 to 4 seeds per berry. The seedless berry is called apirena berry. Externally there are three different areas: beak, belly and back. Inside we find the albumen and embryo.

2.6. Classification of grape varieties.

Meraz, (2013), considers that the Vitacea family has 15 botanical genera, the most important for its commercial value being *Vitis*, from which 110 species are derived. There are approximately 8,000 varieties of grapes in the world, of which a minimal part is used for commercial purposes. The varieties can be classified according to their color into: White, black, red or pink and according to their use in: table grapes, wines and raisins.

Weaver, (1985), mentions that the genus *Vitis* belongs to the family of the Vitaceae, order of the Phanerogams, subtype of the Angiosperms, where all European, American and Asian vines are included. Other researchers have also given them the name of Ampelidaceae, which is the origin of the name given to the description and classification of the different species, hybrids and varieties produced by the crossbreeding of vines and is generally known as Ampelography. It mentions that the varieties are classified as follows: By their botanical characteristics, this classification is based on the description of leaves, branches or clusters which is called Ampelography. By their distribution or geographical origin; French, German, Spanish, American, etc. varieties, when limited to the viticultural geography by nation or by natural regions. By the interest of the destination of the production.

2.6.1. Table grapes.

Weaver, (1985), mentions that they are used for food and decorative purposes. They must have an attractive appearance, good taste qualities, suitable qualities for transport and storage, and resistance to damage incurred in handling.

2.6.2. Raisin Grapes.

Any dried grapes can be included in the raisin designation, although for adequate raisins, the raisins must be smooth in texture and not stick together when stored. Early ripening is important so that the raisins can be removed with considerable time, seedless grapes are preferred and must have a good flavor when dried.

2.6.3. Grapes for juice.

In the production of sweet, unfermented juice, the clarification and preservation process must not destroy the natural flavor of the grape.

2.6.4. Grapes for canning.

Only seedless grapes are suitable for use as canned fruit.

2.6.5. U vas for wine.

Weaver, (1985) and Galet, (1985), mention that these varieties grown in the aforementioned regions can produce satisfactory wines in certain localities. For dry or table wines, grapes with high acidity and moderate sugar content are desirable, while for sweet or dessert wines, grapes with high sugar content and moderately low acidity are required. For mechanical harvesting, grapes should have berries that detach easily from the peduncles. It is clear that this classification is not rigorous, since certain varieties can be used for various destinations, depending mainly on economic circumstances.

2.7. Main wine grape varieties grown in Mexico.

Cetto, (2007), refers to the fact that Mexico currently exports wine to 30 countries, among which the following stand out: England, Germany, France, Holland, Spain, Italy, Canada, United States, even more distant countries such as: Lithuania, Estonia, Russia and Poland. The most important wine-producing states are: Baja California Norte, Chihuahua, Coahuila, Zacatecas, Aguascalientes, Querétaro and Guanajuato.

Cetto, (2007), describes, Reds: Pinot Noir, Cabernet-sauvignon, Merlot, Grenache, Carignan, Salvador, Alicante, Barbera, Zinfandel, Mission, Shiraz, Cabernet Franc, etc. Whites: Ungi Blanc, Chenin Blac, Riesling, Palomino, Verdona, Feher Zagos, Malaga, Colombard, Chardonnay, among others. Vine cultivation in the Cauca Valley is growing. It develops and produces well from 900 to 1600 meters above sea level, although depending on other climatic conditions, it can adapt from sea level to 2100 meters above sea level. The vine adapts to regions with a wide range of temperatures. The winegrowing area of the Valley has an average temperature of 24oC and a wide temperature variation between day and night, which favors the accumulation of sweeteners. It is recommended that this factor be less than 800 mm of rainfall per year, in well-marked periods, in order to reduce the risk of diseases, but it is necessary to have sufficient irrigation for the growth and production of the crop; in the wine-growing area of the Valley there are between 1000 and 1200 mm of rain per year, poorly distributed, which causes fungal problems to the crop.

2.7.1. Shiraz variety.

2.7.1.1. Origin and synonymy.

Galet, (1985), considers that nothing precise is known about its origin, for some it may originate from the city of Shiraz in Persia. The plants were moved from that country by a hermit who planted them in Bessas in the thirteenth century but the first introductions of this variety to France were in the third century when the

Emperor Probus allowed the planting of vines in Gaule. For others, the history of Shiraz would come from the Villa of the city of Syracuse in Sicily, which would explain the different synonyms. The synonyms are: Schiras, Sirac, Syra, Syrac, Syrah, Sirah, Shiraz, also called Petite Syrah.

2.7.1.2. D escription.

The bunches are medium-sized, cylindrical, sometimes winged, compact, with rapidly lignified peduncles, ovoid, small, blue-black grapes, with abundant purine, thin skin but quite resistant and juicy with a pleasant taste Figure 1.

Figure 1.- Shiraz variety.

2.7.1.3. Skills.

Galet, (1988), mentions that it is a late budding, medium vigor and low fertility, so it requires long pruning. Low yields are required to obtain quality wines. In the South of France, this variety has recently been introduced as a quality-improving variety. Clonal selection has been made to improve both production and wine quality. This variety is sensitive to drought and bunch rot (*Botrytis cinerea*). Shiraz is grown in different parts of the world (mainly France, Italy, Greece, Brazil, South Africa and Australia).

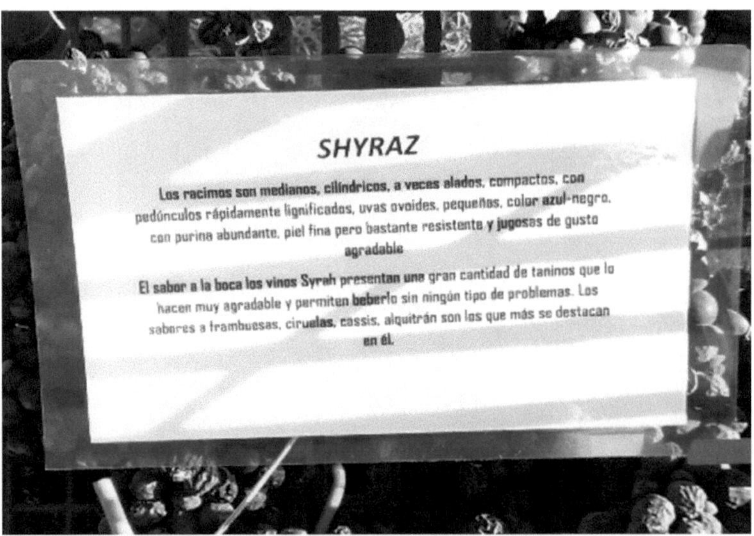

Figure 2. Characteristics of the Shyraz variety.

2.7.2. Merlot variety.

Synonyms: Merlau, Bigney rouge, Vitraille, Plant Medoc. Ampelographically, its growing tip is open, not very hairy and without marked pigmentation, which appears slightly on the internodes. The adult leaves are medium sized, large, with a very dark upper side, with a recurved lobe, sometimes with a tooth at the bottom, with a non-hairy underside and with very little hairiness on the veins, with an open and wide U-shaped petiolar sinus, with wide teeth and straight sides. Merlot is a Bordeaux variety, which spread rapidly in the United States (California) and Mexico because it produces soft red wines. These can be drunk younger; its production is much higher than that of Cabernet Sauvignon, its budding is early (it is done the first week of April in the south of France), this makes it a little more sensitive to late frosts; its maturity occurs in the second season. In autumn its foliage partially thickens; it has yields of 80 hl/ha. It produces soft wines of excellent quality.

In France and Mexico, this variety is blended with Cabernet Sauvignon to obtain a wine that has a good cellaring, finesse, bouquet and beautiful color. To achieve this, in the famous vineyards of Saint Emilion (Bordeaux) they use Merlot, Cabernet Sauvignon and Malbec, at the rate of one third for each cultivar. Cluster of small size, sometimes medium when elongated, of low compactness, with small, somewhat elliptical and distally widened berries, very dark epidermis, with much pruina and very thick, with consistent and quite juicy pulp with particular and very pleasant aromas and flavors, (Galet, 1979; Salazar and Melgarejo, 2005). Figure 2.

Figure 3.- Merlot variety.

2.7.2.1. Maturation .

Salazar and Melgarejo (2005) and Galet, (1990), mention that Merlot can be drunk young, even recently elaborated, they do not require aging in bottle, although its maturation can improve them and make them more complex. As a varietal it gives a wine of rapid evolution, with fresh and fruity aromas and elegant body; to be consumed as a young red wine or as a young wine with a few months in oak barrels. Authentic red Bordeaux cultivar, of high vigor with a tendency to very abundant branching and upright growth; of good fertility but low production, early voting, therefore sensitive to spring frosts, and also to winter frosts. It is sensitive to bunch runner in limiting climatic conditions. It requires short pruning, is sensitive to mildew, downy mildew, green mosquito, and does not tolerate poor, dry soils where it shows a clear tendency to flower drop. It is the basis for very round and complex wines with excellent color and alcohol content, tannic and smooth at the same time, very suitable for aging. Today it is considered one of the best varieties for cultivation, with high phytoalexin content and therefore with a certain resistance to various pathologies. Due to its high sensitivity to phylloxera, it should be grafted on resistant rootstocks, the most common are mainly: SO-4, 420-A, Riparia Gloria, 161-49.

Figure 4. Characteristics of the Merlot variety.

2.7.3. Cabernet Sauvignon Varietal.

Weaver, (1976) and Winkler, (1970), mention that the variety is quite homogeneous, with some differences in the shape of the cluster and in the typical characteristics of the wine. It is a fairly vigorous variety with medium-late budding, fairly erect vegetation and medium-short internodes. It adapts to temperate climates and is best in dry or well ventilated areas, in the north it prefers areas well exposed to the sun, on hills and light soils, especially in the valleys. It does not accept excessively fertile and humid soils that induce great vigor and lignification difficulties. It adapts well to different pruning rules taking into account the pedo-climatic conditions. Production is regular and constant, maturing in the third season. Resistance to diseases is normal, but it can be considered somewhat sensitive to bunch drying, so it is necessary to take into account the K/Mg ratio of the soil. Important in the production of the famous wines of the Gironde region of France. In appropriate California locations, this grape produces a wine with a pronounced varietal flavor, high acidity and good color. It is one of the best varieties for red wine making. Clusters are small to medium, irregularly shaped but often conical - long, sparse to well-filled. Berries are small, many-seeded, almost spherical and black, with gray pruina. They ripen in mid-season. The skin is strong and the flavor is pronounced and characteristic. The vines are very vigorous and productive with cane pruning. These grapes reach their highest quality in the cooler parts of the coastal valleys.

Figure 5.- Cabernet Sauvignon variety.

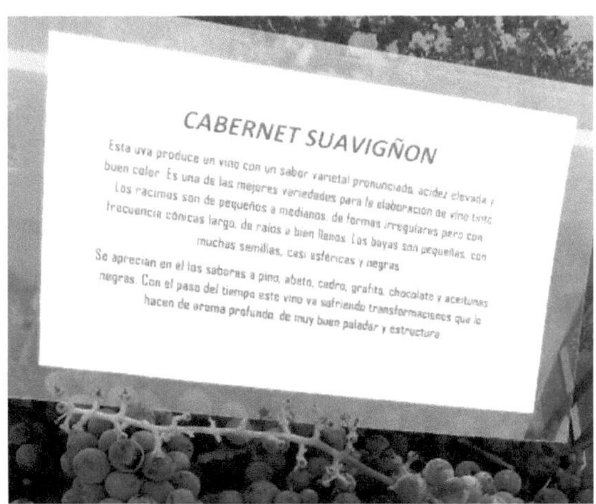

Figure 6. Characteristics of the Cabernet Sauvignon variety.

2.7.4. Tempranillo variety.

Martínez (1991), considers that this variety had an important boom in our country from the 1950s onwards. In the 1936 census, this grape variety is not mentioned, possibly grouped as "various red French varieties" (Junta Reguladora de Vinos, 1936). Afterwards, it began to be planted, taking advantage of its excellent yields. And so it is that in the 1968 census 10,916 hectares appear, almost entirely located in the different departments of Mendoza, Spanish Tempranillo and is characterized by large, pentalobed leaves, with well-marked lateral sinuses and with the petiole generally perforated. The underside of the leaf has abundant "spider web". The cluster is conical, elongated, well filled, large, with reddish black berries, spherical and medium (1.6 cm. long). There are a multitude of clones.

Figure 7.- Tempranillo variety.

2.7.4.1. Evolution of planted area.

Likewise, in GIV (2015), it is mentioned that subsequently, all the red grape varieties suffered the same fate and about 50% of the planted area was uprooted. The 2000 census cites 4,335 hectares located in the different winegrowing regions of Mendoza. By 2005, the planted area grew to 6,099 hectares.

2.7.4.2. Viticultural characteristics .

It is mentioned that the grapes ripen in mid-March. Traditionally, the grapes have been grown in vineyards, seeking to take advantage of their productive potential, but it is precisely with high yields that their quality decreases, resulting in, among other characteristics, a loss of color in the wine and the presence of green and drying tannins.

2.7.4.3. Strain bearing: upright .

Leaf characteristics: Large size, pentagonal shape, very deep lateral sinuses, seven-lobed leaves, dark green beam, almost black and velvety underside, young pampas have cottony budding, with reddish pigmentation at the end. Cluster characteristics: The cluster is conical, elongated, well filled, large. Berry characteristics: Medium size, spherical shape and blue-black color (1.6 cm. long). Vegetative period: Late to medium sprouting and medium-early maturity. Yield: 2.5 to 3.2 kg/straw.

2.7.4.4. Organoleptic characteristics.

It is said that this grape variety, in its youth, exhibits simple fruity aromas with aromatic notes of blackberry, raspberry, blackberry, plum, fig and also raisins, which would indicate a notable amount of aromatic percussors, but the aged specimens mature framed in envelopes.

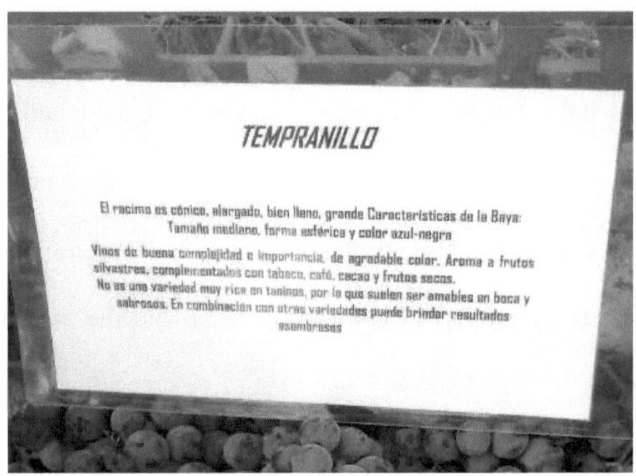

Figure 8. Characteristics of the Tempranillo variety.

2.7.5. Malbec variety.

Rodríguez, (1996), mentions that it is the most important red grape variety in Argentine viticulture, especially in Mendoza, where more than 80% of the national total is produced. The areas located at higher altitudes are experiencing an increase in the cultivated area, since they produce grapes with a high content of phenolic compounds, responsible for many of the desired characteristics in red wines, due to their organoleptic value and antioxidant capacity. The amount of ultraviolet-B radiation (UV-B; 280-315 nm) reaching the earth's surface increases with altitude. In Mendoza, the highest vineyards, around 1,500 meters above sea level, receive UV-B levels that can increase the phenol content in leaves and berries, although they could also be harmful to the normal growth and development of the different tissues.

Figure 9.- Malbec variety.

Abscisic acid (ABA) is a phytohormone that mediates plant responses to different stress factors (salinity, temperature and water deficiency), and possibly also to UV-B, mitigating their negative effects. It is considered the emblematic red grape and is characterized by being a demanding cultivar (cv.), with respect to the agroecological conditions of the production areas and vineyard management, at the moment of expressing its full potential. According to several authors, it originates from the areas of Cahors, Quercy or Touraine, in the southwest of France. There it is known mainly by the name "Côt", although it has several synonyms such as Auxerrois, Côt de Bordeaux, Pressacn and Luckens, among others. The name "Malbec", by which it is known in Argentina, comes from "Malbeck", the name of the Hungarian winegrower who introduced this variety in Bordeaux.

It was widely cultivated in French vineyards, representing a large part of the vineyards in the south-west and center-west of that country. It was an important variety until the time of phylloxera, where it was badly affected, being relegated until its reintroduction in 1940. In the Cahors region it is the dominant red variety, and according to the Appellation d'Origine Contrôlée (AOC) it is at least 70% of the wines produced, complemented by other varieties such as Tannat and Merlot. It is also part of the AOC "Bordeaux" red wine blend, together with Cabernet Sauvignon, Cabernet Franc, Carmenere, Merlot and Petit Verdot.

2.7.5.1. Characteristics of the Malbec variety.

To the eye it presents colors so intense and dark that in certain cases they can be confused with black, these colors are defined by cherry red or cherry red. The main aromas of this wine are made up of cherries, plums, coffee, chocolate, leather, truffle, vanilla, among others. The primary aromas are mainly truffle and vanilla. On the palate, Malbec wines appear with flavors of plum jam, cherry jam, chocolate, dried fruits, vanilla and balsamic tastes. This variety of wines are warm, smooth and with a certain amount of sweet tannins that characterize them in a very pleasant way.

2.8. Evaluation of Grapevine Varieties in Northern Mexico.

Rodríguez Vázquez, (2018), in the first stage of evaluation in 2017 in the town of Monterrey Durango, in a randomized complete block experimental design with five treatments (wine grape varieties) and five replications, obtaining 25 experimental units. The total experimental area was 10 m long and 15 m wide, obtaining 150 m^2. The useful experimental plot area consisted of a line 2.0 m wide and 10.0 m long, obtaining an area of 20 m^2. The variables evaluated were weight per bunch, number of berries per bunch, number of bunches per plant, experimental yield (kg/m2), commercial yield (t ha-1), berry volume, fruit pH and soluble solids content (° Brix). The results showed that for weight per bunch and number of grapes per bunch, treatment 2 (cv Merlot) was the best. For soluble solids content in Brix degrees at seven days before and seven days after harvest, the best treatment was found to be Treatment 1 (Cabernet Sauvignon cultivar), with a value of 25 and 24.2 ° Brix. For fruit pH, Treatment 3 (cv Malbec) had the highest berry pH value of 4.6. In the volume per 10 berries, Treatment 1 (Cabernet Sauvignon cultivar) had the highest value of 8.6. Finally, in the volume per bunch of grapes, Treatment 3 (cv Malbec) presented the highest value with 137.54 cm3. The objective of this study was to evaluate the phenological and productive behavior of five grape varieties intended for red wine production.

On the other hand López Aguilar, (2018), in his second evaluation found the following results in the same locality; the evaluated variables stem perimeter, stem diameter, number of bunches per plant, number of total berries per harvested bunches, number of bunches per hectare, weight per berry, weight of 10 berries, weight per bunch, total weight of bunches per plant, tons per hectare, volume of 10 berries, sugar content, pH of the fruit. The results showed that Treatment 1 (cv Cabernet) excelled in stem perimeter with 16.84 cm, stem diameter with 8.42 cm, number of clusters per plant with 40 clusters and number of clusters per hectare with 13,444 clusters, respectively. Treatment 2 (cv Merlot) was the best in the number of total berries per bunches harvested with 2671.7 berries, in total weight of bunches per plant with 3.48 kg and in tons per hectare with 1.16 tons. Treatment 3 (cv Shyraz) excelled in weight per berry with 2.2 g, in weight of 10 berries with 22.0 grams, in volume of 10 berries with 19.33 cm3 and in sugar content with 27.46 degrees Brix. Finally, Treatment 5 (cv Tempranillo), excelled in weight per bunch with 163.16 g.

For 2019, with the same treatments and repetitions and the five cultivars (cv Cabernet, cv Merlot, cv Malbec, cv Shyraz, cv Tempranillo), and where the variables evaluated were stem perimeter, stem diameter, number of bunches per plant, number of total berries per bunches harvested, number of bunches per hectare, weight per berry, weight of 10 berries, weight per bunch, total weight of bunches per plant, tons per hectare, volume of 10 berries, sugar content, pH of the fruit, and the following was found: According to the data analyzed, the Tempanillo variety was the one that stood out in the variables under study, followed by Shyraz, then Merlot, Malbec and finally Cabernet Suavigñon, however it is important to highlight that the most important variable would be Cluster Weight, and Brix Degrees reached, with Tempanillo and Merlot being the most outstanding.

We can conclude that the Shiraz, Merlot, Cabernet Sauvignon, Tempranillo and Malbec (*Vitis vinifera* L.) are the best varieties in the order indicated, which are adapted to the areas of Mexico. They excel in grape production variables as well as in quality.

2.9. Characteristics of wine grapes .

Weaver, (1985), mentions that wine grapes are harvested by hand or with mechanical harvesters. The appropriate time for them depends mainly on the type of wine to be made. Dry wine grapes should have high acidity and moderate sugar content. Therefore, they are harvested when they have 20 to 24° Brix. Those grapes destined for sweet wines should have as high a sugar content as possible and moderate acidity, without becoming raisiny, with a graduation of 24°Brix or higher.

2.9.1. Practices to Improve Grape Quality

Archer and Strauss, (1985), refer that one of the important practices is the management of foliage, since it plays an important role in the vine plant. Therefore, and from this perspective, canopy promotion cannot be restricted only to the plant itself, but to each and every one of the direct or indirect aspects that influence its physical appearance and yield. The importance of canopy promotion has been increasing from being an integral practice, absolutely essential in viticulture and enology in order to obtain and improve grape and wine quality. The ultimate goal of plant promotion is to obtain homogeneous canopy, which carries out photosynthesis efficiently, forming shoots of similar vigor and uniformly distributed that produce healthy and high quality grapes, with similar clusters, of similar berry size and uniform maturity. In addition, to maintain longevity, the growth and development of other parts of the plant should not be affected.

2.9.2. Procedure for the production of artisanal or homemade red wine.

Morales-Salazar, A. 2020. Personal Communication.

- Grape Harvest; to determine the time of harvest for these grapes it is important to determine the sugar content in the grapes, this is done by determining the Brix degrees (^{O}Bx) with a refractometer at least every week from the second half of July, once it reaches an average of 23 ^{O}Bx it is time to harvest and once harvested it is important to wash the bunches with running water.

- Once the grapes are harvested, we proceed to remove the berries from the bunches, this process is called destemming, discarding the hard structures of the cluster, this is preferably done cluster by cluster by cluster by hand, this when there are low quantities of grapes of about 50 kg, in the wine industry regularly has industrial equipment for this stage of the process.

- Sanitization; it is important that all utensils used for these processes are perfectly clean, do not wash them with soap, only with running water and rinse with Sodium Metabisulfite (MBSNa), ($Na_2S_2O_5$) this product is a disinfectant, antioxidant and preservative.

- Maceration; once the grapes are harvested, they are macerated with the hands, preferably leaving the juice and the must that corresponds to the skins or skin and the seed, trying to extract as much juice as possible, this is stored in wide mouth containers to once put them at two thirds of their volume to start the application of products for fermentation.

- Fermentation process is considered the process in which ethyl alcohol is generated and carbon dioxide is released from the process of decomposition of the must and the alcohol is deposited in the grape juice, product of the maceration of the grape bunch, which has been deposited in a wide spout container, and where the following products are applied to help this process;

- Potassium metabisulfite (MBSk), ($K_2S_2O_5$). Antioxidant and preservative and fungicide product, which helps to extract the color and regulates the successive fermentation in cases of very high environmental temperatures, where environmental temperatures exceed 30 °C. The recommended dose is 1 g per 10 l of juice plus must. It should be dissolved before application in 150 to 200 ml of water at room temperature and in case it is not available, it should be heated in a water bath until it reaches 28 to 30 °C.

- Yeast (*Saccharomyces cerevisiae*), or bread yeast, helping the fermentation process, the recommended dose is 2 g per liter of juice plus must. In the case of the yeast, the dissolution will be in hot water of 35 to 36 °C, preferably in glass containers, do not use pewter containers, the containers to be used should be sanitized with MBSNa, it can also be dissolved in grape juice at room temperature.

- Nutrient; they are usually products derived from carbohydrates, which the bacteria contained in the yeast feed on, to carry out their vital functions to obtain energy and continue the fermentation process.

- Conservation temperature, throughout the process from the maceration until it is bottled and even in the storage process of red wine it is important to maintain the temperature in a range of 10 to 30 °C if the environmental conditions do not allow it, it is recommended to put the containers in a water bath with ice and monitor the temperature that does not go outside the recommended range is to monitor with a thermometer manually.

The fermentation process varies according to the ambient temperature, type of juice and handling of the process, it can last from 3 to 7 days, which has to take care of the recommended temperature of the juice (10 to 30 °C). The juice and the must, which are in wide mouth containers, are covered with a piece of cloth, preferably cotton, which are filled to two thirds of their capacity because the fermentation process causes excess volume, forming a dense layer on top which is called "Sombrero". It is recommended to shake each container in the mornings and evenings, this can be done manually or with a plastic or stainless steel utensil, and in each movement the temperature of the juice is taken, and the Brix degrees (°BX) are determined, in case of not having a refractometer which is the instrument that determines Brix degrees, it is determined visually or by touch, this is when when the juice is no longer bubbling when you put your hand in it. This is when the fermentation process is finished. If the instrument is available, the reading is taken twice a day, in the morning and in the afternoon, and when the refractometer reads 0 °BX the fermentation process is finished.

Once this happens the so-called Sombrero is taken by a piece of cloth and the juice is filtered into the container, the filtered Sombrero is discarded.

Whenever the juice and must are stirred, it is recommended to apply MBSK at the recommended dosage, and it is very important to maintain the proper temperature.

Once the fermentation process has finished in its first phase, which can vary from 4 to 7 days, depending on the handling of the product, the juice is deposited in plastic jugs previously sanitized with MBSNa, covering and sealing them hermetically with tape. In the following weeks or months, as the case may be, the demijohns are uncorked for 4 to 5 seconds daily for 10 to 15 days, since the final fermentation process involves the release of CO_2. The demijohns are opened and closed for short periods of time to prevent the entry of oxygen (O_2), which is detrimental to the proper development of the wine,

After 15 days it can last from 2 to 3 months in the demijohns and during this time a pinch of MBSk will be added every third day. After these 3 months, which could be up to 7 months, the tape is removed and the demijohn is uncovered and it is recommended to make two or three rackings twice, removing the sediments that still remain in the grape juice of the demijohn.

After all this process, if the winemaker so wishes, the wine can be transferred to amber bottles and closed with corks, which do not allow oxygen to enter, as this is harmful to the conservation of the wine.

2.10. Management of the main cultivation practices that affect the yield and quality of grapes for wine production.

2.10.1. Physiological and practical considerations.

It is necessary to briefly mention the role played by aspects such as planting density, trellis type and water management. In order to obtain a growth that avoids an excess of shoots and achieves optimal levels of water consumption and soil utilization by the roots. It is recommended to apply high planting density and smaller trellises in soils with low to medium potential, while lower densities and larger trellises can be used in soils with medium to high potential.

Water table and irrigation water availability will affect density in both scenarios. The choice of trellis system is a function of soil potential, vigor of the variety and rootstock combination, climate, mechanical practices and maintenance requirements. Although numerous trellising systems are used. A balanced vine with photosynthetically efficient foliage should always be sought. It is advisable to control growth so that there is not an excess of shoots, the interior shade of the foliage is limited, and there is sufficient space for the shoots to reach a minimum of 1.4 m or support about 16 primary leaves. Therefore, to increase grape quality and decrease production costs, trellis systems should be governed by basic canopy management principles. In summer, when normal daytime temperatures are outside the ideal range for optimum berry coloration (15-25° C) there is the possibility of an increase in pH, grape size takes on great importance as a potential quality parameter, due to the higher skin/pulp ratio and the greater capacity for extraction of phenolic compounds (especially anthocyanins) in the smaller berries. Under such conditions, the practice of irrigation during the stage of cell division in grapes should aim at reducing grape size.

Despite the marked resistance of this parameter during the ripening period, it is sensitive to water stress, improved lighting conditions, and competition with vegetative growth before veraison. Grapes, despite depending on primary precursors (such as sucrose and amino acids) from leaves, are also metabolically active in the formation of secondary compounds such as isoprenoids involved in aroma (monoterpenes) and nitrogenous compounds such as 2-methoxy-3-isobutypyrazine, responsible for the typical fat and green pepper aroma of Sauvignonbla, Cabernet-sauvignon and Sémillon varieties (Archer and Strauss, 1985; Archer, 1988; Lacey et al., 1991; Greenspan, 1994 and Carbonneau and Cargnello, 1999).

2.10.2. Pests and diseases.

2.10.2.1. Phylloxera *vastatrix (Phylloxera vastatrix P.)*

Hidalgo, (1975) and Weaver, (1985), refer to the phylloxera as an aphid, which belongs to the Aphidae family (order: Homoptera); they are sucking insects and their color is variable, yellow, red, green, gray, black. This pest is the most important pest of grapevines. It is a 1 mm long aphid that lives on the roots, from which it absorbs the sap and facilitates the entry of fungi that kill the roots, causing the death of the plant. In ungrafted *Vitis vinifera* vineyards, phylloxera is manifested by the appearance of areas of weakened plants without apparent causes, this general weakening of the plants is a consequence of the disorganization of the root system of the vine, due to the bites of phylloxera to nourish itself at the expense of the sap. The holes caused by the aphid in the rootlets favor the rotting of these organs and, as a consequence, the vine is weakened, taking on a stunted appearance and producing shoots with short internodes and small, yellowish leaves, eventually drying out and dying after a few years.

This insect is the most widely known aphid due to the destruction of vineyards it has caused worldwide. It is native to the USA, west of the Rocky Mountains, and the first European country to suffer its effects was England in 1863. In terms of climate, it is said that agricultural phylloxera outbreaks are greater in dry years with high temperatures, partly because the insect attacks in spring are earlier, thus affecting a large number of young growing leaves, and partly because the succession of generations is more rapid, thus increasing the population of the insects. On the other hand, in cool and rainy springs, galls rarely form.

Galet (1979) states that the use of phylloxera-resistant rootstocks is necessary in practically all soils, but can only be dispensed with in sandy soils where this insect cannot carry out its invasion, since its mobility there is very reduced.

2.10.2.1.1. Biological cycle of phylloxera.

Ferraro, (1984), and Ruiz (2000), mention that the females of the so-called sexed generation lay the winter eggs (only one per female) on the bark of the vines, in two or three years, coinciding with the sprouting of the plant, the hens hatch and settle on the leaves, founding the first colonies. The adult females are apterous and reproduce by parthenogenesis. The foundress lays about 500 eggs inside the gill during one month. After 8-10 days they hatch and the neogallicolous gall females appear, these migrate from the gill and form new colonies in successive gall generations by parthenogenesis. An ever-growing part of the gall larvae leaves the leaves to go to the roots, where they form colonies of neogallicolae-radicolae developing several generations during the summer also by parthenogenesis. On the other hand Perez, (2015), says that at the end of the summer appear the winged sexiparous females that go outside and lay eggs on the vine shoots, but some will give rise to males and others to females, forming the generation called sexed. The fertilized female is in charge of laying the winter egg. This closes the cycle.

2.10.2.1.2. Symptoms of phylloxera damage.

Ferraro, (1984), and Madero, (1997), refer that phylloxera in vineyards is manifested by the appearance of weak plants with no apparent cause. This weakness gradually spreads, forming an attacked area in the form of a round spot, which expands in concentric circles.

The phylloxera bite to the root, apart from deforming its growth, causes the entry of fungi and rotting of the roots, causing the weakening and death of the plant. Depending on the age of the roots, this insect produces two types of lesions: Salazar and Melgarejo, (2005) and Ruiz, (2000) consider the following points:

a) Knotweed: this is in roots that have not developed epidermis, which make it lose vitality, which arise as a result of the bite of the parasite on the end of the rootlets of the vine, which are in full growth, the insect introduces its stylet to the phloem to suck the sap, the next day the injured rootlets change their cylindrical shape to another bulging, bright yellow, two days later gives rise to a knotty which reaches its final size in the next 10 or 15 days.

b) Tuberosities: as the epidermis is fully developed, formed in the thickest roots by the action of the insect, the wound is caused by the insect's stylet and has no action on the cambium; however, on the surface of the root, which surrounds the wound, there are irregularly shaped bulges that give the organ a wavy shape.

In European rootstocks, the classic symptoms of root diseases (stunted vegetation, chlorosis) are observed. In the root system, the larvae's feeding bites produce a hypertrophy in the rootlets (knotweed), which, when decomposing, determine the progressive destruction of the root system. In American grapevines (mother plants) a strong attack on the leaves (galls) can cause a decrease in growth and a bad wood setting.

Ferraro, (1984). He mentions that phylloxera can spread actively by the insect, or passively, with the intervention of man, depending on the conditions of the environment, climate, soil, variety of vine cultivated and the type of phylloxera in its evolution. The general weakening of the plants appears as a consequence of the disorganization of the root system of the vine, because the bites that the insect makes in the root to suck the sap, favor the putrefaction of these organs, preventing the sap to continue its normal course towards the aerial part of the plant.

2.10.2.1.3. Phylloxera control methods.

Reynier, (2001), mentions that V. vinifera varieties (Merlot, Cabernet-Sauvignon, among others) offer practically no resistance to phylloxera attack, nematodes and Texas rot, to which a score of 1/20 can be given, while the American species, thanks to the rapid formation of a layer of healing sap, present a resistance that can be between 16/20 and 18/20. Gallicolous generations are sometimes detrimental to the cultivation of rootstock mother-rootstocks and the production of rooted rootstock plants.

- Soil treatment with carbon disulfide or DDT, in the dichloroethyl ether state, kills many of the insects, but these treatments are very expensive and must be repeated frequently (Winkler, 1970).

- Prolonged waterlogging of the soil in mid-winter kills many insects, but larvae may survive for up to three months.

- More than a century of experience has shown that grafting V. vinifera varieties onto resistant rootstocks is a safe and permanent means of protection against phylloxera, provided that a sufficiently resistant rootstock is used.

There is a range of rootstocks adapted to different types of soil and obtained mainly from the species V. riparia, V. rupestris and V. berlandieri, which offer a sufficient guarantee.

2.10.2.2. Nematodes.

Martínez et al, (1990) and Hidalgo, (1975), consider that the presence of nematodes is another factor to be taken into account when choosing the rootstock.

The main nematodes that attack grapevines are classified into two groups:

a). Ectoparasites: are those that live in the soil extracting nutrients from the roots, but without penetrating them.

b). Endoparasites: are those that penetrate entirely into the roots where they live, feed, grow and reproduce.

The former do not cause significant direct damage; on the other hand, some play a fundamental role in the transmission of specific grapevine viruses; such is the case of the genus Xyphinema.

Of the endoparasitic nematodes, the two most important genera are:

a). Meloidogyne: includes the most damaging endoparasitic nematodes for grapevines. They develop mainly in light, sandy soils; they are widespread in the vineyards of California (U.S.A.) and Australia, where they cause significant damage. The larvae of this type of nematode penetrate into the young roots through the coppice or piloriza.

b). Pratylenchus: These nematodes have migratory habits and cause necrosis, infect other roots and so on until the life of the vine is compromised. This whole process is aided by soil microorganisms that settle in the roots causing rotting and disintegration.

Among the ectoparasitic, virus-transmitting nematodes is the genus Xyphinema, which should also be taken into account. The strongest nematode pest in grapevine is *Meloidogyne incognita var.* Acritachitwood. The damage it causes is similar to that caused by phylloxera; it causes abnormal cell growth, characterized by galls or swellings in the form of a collar on the roots, while those caused by phylloxera are only observed on one side of the root.

Some nematode resistant rootstocks are: Dog Ridge, Salt Creek, 99-R (very resistant): 110-R, 140-Ru, Rupestris de Lot, 420-AM, among others.

2.10.2.2.1. Symptoms of nematode damage.

Winkler,(1970), refers to the fact that it is usually difficult to identify when a plantation is attacked by nematodes, because they live underground and cannot be seen with the naked eye. In general, they can be observed:

- Weak plants, with little development and high susceptibility to attack by other pests or diseases.

•. Root nematodes cause abnormal cell growth resulting in characteristic tumors. In young rootlets, galls appear as widenings of the entire root that manifest as a series of knots resembling a string of beads, or the swellings may be so close together as to cause a rough continuous thickening of the rootlet over a length of 2.5 cm or more.

2.10.2.2.2. Nematode control methods.

Chávez and Arata, (2004) and Rodríguez, (1996), mention that to prevent and combat nematodes we must:

Use rootstocks or rootstocks of American vines with resistance to nematodes. V. berlandieri or V. riparia, on which the varieties are grafted.

The use of manure in fertilization practices does not allow the proliferation of nematodes, because it contains fungi and other natural enemies of nematodes.

Encourage the existence of earthworms, their excreta are toxic to nematodes.

As an extreme measure due to their high toxicity, the use of nematicides: Aldicarb (Temik): Oxamyl (Vidate): Carbofuran (Furadan) among others.

In this case, it must be taken into account that nematicides leave toxic residues on plants and affect consumers for very long periods of time, in some cases up to 10 years.

2.10.2.3. Texan rot (*Phymatotrichum omnivorum*).

Vargas *et al.* (2006) and Anonymous (1988) mention that among the root pathogens that affect soil productivity is *Phymatotrichum omnivorum*, the causal agent of root rot or Texan rot, a disease of economic importance, both for its effects on production and for its wide distribution in agricultural regions of Sonora, Chihuahua, Coahuila and Durango. Omnivorum proliferates rapidly in calcareous soils of northern Mexico and southwestern United States.

The damage caused to the roots results in symptoms on the foliage of the attacked plant, which generally occur from late May and early June until October, when conditions are ripe for the development of the pathogen. Sometimes, in young plants the symptoms advance very quickly, since they wilt suddenly without having presented any symptoms in previous days. In these cases the dry leaves remain attached to the plant for some time. In adult vines the leaves often show yellowish spots at the beginning, later in the same year or in the following years, the plants lose vigor, the leaves dry up and fall, leaving the vine partially or totally defoliated.

2.10.2.3.1. Control methods.

Mortensen, (1939) and Castrejón (1975), point out that based on the above and knowing the devastating effects of this fungus, it has become necessary the possibility of rootstocks tolerant to this disease. In studies carried out in Texas, U. S. A. for several years, it has been possible to detect resistance in the species Vitis candicans, Vitis berlandieri being these native to northern Mexico, indicating that the rootstocks Dog Ridge, Salt Creek and Teleki 5-C, tolerate the fungus.

2.11. Rootstocks in grapevine cultivation tolerant to pests, diseases and soils.

2.11.1. Origin of the grafts.

Salazar and Melgarejo, (2005) mention that the origins of the rootstocks are pure American species such as Vitis riparia and V. rupestris, planted directly. Hybrids of V. riparia with V. rupestris. The American species V. berlandieri, resistant to limestone, was hybridized with V. vinifera, V. riparia and V. rupestris. Use of V. solonis, found in America, in saline soil. Complex hybrids with intervention of these and other species.

2.11.2. Use of rootstocks

Boulay, (1965) and Madero, (1997), consider that the main objective of using rootstocks is to combat phylloxera and nematodes, on the other hand it helps cultivars to adapt equally to different soil and climatic conditions or resistance to pests and diseases, resorting in these cases to rootstocks capable of withstanding the soil conditions and which in turn are compatible with the variety. It is also undoubtedly the most effective and affordable method commonly used in vineyards worldwide to control the damage caused by phylloxera, and also to deal with other problems that are present in the region's soils, such as nematodes and diseases like Texas rot.

2.11.3. Advantages of the use of rootstocks.

Calderón, (1977) and Boulay, (1965), refer that among the main adverse factors that the rootstock can be resistant to are: presence of various types of pathogens such as pests, nematodes and diseases, salts, alkalinity, excess limestone, poor drainage, excess humidity, drought, etc. The behavior of the rootstocks plays a very important role, since the correct choice of these will depend to a great extent on the production of the orchard, since the rootstock will act in combination with the graft in relation to the environment. It must be taken into account that there is no universal rootstock, it must be taken into account the growing environment, soil, climate, the species and variety to be grown, the compatibility of the necessary graft, parasitic sensitivity, etc., the relationship of a weak rootstock with a vigorous and reciprocal rootstock.

2.11.4. Characteristics of a good rootstock .

Agusti, (2004), mentions that there is no ideal rootstock, since there are many factors that influence its behavior and, above all, its relationship with the grafted variety. However, it is possible to establish a series of characteristics that, in general terms, define its quality, although they are subject to variations induced by the environment.

Since the use of rootstocks is mandatory as a practical and efficient solution to deal with phylloxera, when selecting the most appropriate rootstock for each particular vineyard, it is necessary to consider factors or conditions present in each case. These include the presence of nematodes, texan rot, active limestone, drought, excess humidity, salinity, as well as soil type and depth. The selection of the appropriate rootstock for the problem to be combated is a very important and decisive aspect that deserves our full attention, since this decision, once the vineyard is established, will be carried through all the years of its productive life.

For proper rootstock selection, at least five fundamental conditions must be considered:

- Be resistant to phylloxera.
- Be resistant to Texan Rot and nematodes.
- Show adaptation to the environment, climate and soil.
- Have satisfactory affinity with the producing variety.
- Allow the development of the plants destination of the grapes.

2.11.5. Rootstock selection according to the grafted variety.

Larrea, (1981), mentions that a viticulturist must properly select the rootstock for his vineyard, since the success and longevity of the vineyard depends on it. At present there is no rootstock that can be used for a large region due to the great variety of soils, topography and adaptation of vines. The rootstock is selected taking into account phylloxera resistance, resistance to calcareous soils, drought, salts, nematodes, adaptation to acid soils, vigor, and earliness or delay of harvest.

2.11.6. The quality and vigor of grafts.

Hidalgo, (1975), mentions that it is an accepted norm in viticulture that the obtaining of high quality is opposed to the adoption of any practice that has as a consequence an increase in the vegetative capacity of the plant. In viticultural situations where the aim is to produce quality wines, the choice of rootstock should be oriented towards those with weaker vegetation, naturally compatible with its normal and economical development. On the other hand, in winegrowing situations aimed at the production of ordinary wines, the needs are totally different, requiring a vigorous rootstock for abundant production. Making both concepts compatible, we can summarize by saying that, in environments with a vocation for quality, the most vigorous rootstock should be chosen from among the weakest adapted to the circumstances, while in situations with a vocation for quantity, the rootstock that best adapts to the conditions of the environment should be chosen, with vigorous development, inducing high yields. It should also be taken into account that rootstocks with very vigorous development induce bunch growth of wine grape varieties, which are prone to it, and there is a greater risk of cryptogamic diseases. On the other hand, medium vigor rootstocks, in soils to which they are adapted, give regular and abundant fruiting, with normal production and ripening. Very weak rootstocks should be advised with extreme caution, only for very good and very particular soils, which limit their use.

2.11.7. Incompatibility.

Hartman and Kester, (1979) mention that incompatibility is defined as the inability of two different plants, grafted together, to successfully produce a union and develop satisfactorily as a composite plant. Revealing symptoms of this problem are that the graft union does not take place, or that the union is satisfactory but after some time deformations are generated, excessive growth of one of the two parts, yellowing of the foliage with early defoliation and premature death of the plant.

2.11.8. Influence of rootstocks on plant vigor.

Muñoz, (2002), mentions that it has been determined that in very fertile soils very vigorous rootstocks could cause a decrease in productivity due to an excess of shade to the fruit causing poor quality. In poor soils and lack of moisture, vigorous rootstocks would have a greater capacity to survive, due to a greater penetration of the root system, which would allow a greater absorption of nutrients, thus favoring the vigor of the graft. Considering all this, the choice of a particular rootstock with respect to its vigor, should take into consideration whether the growing conditions are favorable or not, which will be determined by soil fertility, water availability, climatic conditions and plant conduction systems.

On the other hand, he considers that the growth of a vineyard depends on the leaf area, since it is the system that captures light energy, necessary for ripening, growth and accumulation of reserves of compounds in the grape and the vine. The leaf area determines the vineyard's potential as an instrument that captures light energy and transforms it into dry matter; therefore, the greater the leaf mass and the more energy it captures, the greater the development will be. It is then that a conditioning factor arises, and that is that this leads to a dangerous alteration of the microclimate both inside and around the vegetation.

On the other hand, Martínez and Carreño (1991), consider that the vigor of the rootstock, together with that of the variety, determines the vigor of the plant, so this factor influences production, quality, ripening time and even the load of buds left on pruning in general, vigorous rootstocks such as Salt Creek, Dog Ridge, 110-R, 140-Ru, among others, favor high yields, delay ripening and sometimes require a greater load of buds left on pruning to avoid problems of flower runout from the bunch, while weak or medium vigor rootstocks such as 420-A, Teleki-5C, SO-4, tend to favor quality and also advance ripening.

It is well known that rootstocks play an important role on the ripening process and on the final quality of the grapes, influencing mainly by the vigor they confer to the vegetative system, since the most vigorous vineyards are always the least precocious, giving the least sugary and most acidic fruit.

2.11.9. Influence of rootstocks on grape production and quality.

The rootstock can influence the quality of the fruit produced, and it is unlikely that there is a direct influence of the rootstock on quality. Experiences abroad, comparing grapes from grafted vines with fruit from ungrafted plants, indicate that there are differences with fruit from ungrafted plants, there are noticeable differences in sugar content, pH and berry weight. Vigorous rootstocks give higher yields per plant but lower sugar content and produce some delay in ripening. However, excess vigor can sometimes result in poor fruit set, while weak rootstocks give lower production, higher quality and earlier ripening. A characteristic of the rootstock is the production capacity of the variety. In general, the vigor of the rootstock could be related to a low production level of the grafted variety. It has been determined that the production of a variety varies considerably according to the rootstock.

Fruit quantity and quality are two of the points where it has been very difficult to find a clear effect attributable to rootstocks. The results obtained in the different trials have been erratic. Although in some cultivars a higher yield has been observed with a certain rootstock, this cannot be attributed to an improvement in fruit quality (greater diameter and weight), but rather to a greater number of bunches where even quality has been adversely affected. In other cases, when better fruit quality has been observed, quantity has been sacrificed.

The 140-Ru is one of the rootstocks with which good production and berry size are obtained, in addition to increasing sugar content and color in the "Italia" variety. The SO-4 rootstock induces the production of small berries, somewhat compact clusters in the "Italia" variety. The number of clusters per plant has its origin and initial development within the fertile bud. Fertility differs between varieties and is influenced by the vigor of the shoot and rootstock. The presence of one or more clusters in each bud, as well as their size, depends on the growing conditions and the environment, in situations that alter the normal growth cycle of the vine, delaying the initiation of fruiting buds. Berry weight in table grapes is an important aspect of quality. It has been observed that some rootstocks with weak vigor produce an increase in berry weight, while in others it may decrease. It is not yet clear that all the effects on fruit quality are due directly to the rootstock, or are due to the change in the microclimate (Hidalgo, 1975; Martínez *et al.,* 1990; *Martínez and Carreño, 1991; González, 1999; Martínez and Carreño.* 1991; González, 1999; González and Muñoz, 2000; and Ljubetic, 2008).

According to Delgado (2012), based on the data obtained in a research work, he concluded that SO-4 is the most suitable rootstock for the shiraz variety, since with it a higher grape production was obtained (12.1 ton ha-1) without deterioration of grape quality. In the case of sugar accumulation, good, since with SO-4, less concentration was obtained, this is more than sufficient for the desired objective (production of red wine), although it may be that in the case of the other rootstocks the harvest should be earlier.

2.12. Vitis species used to produce rootstocks.

2.12.1. Vitis riparia .

Martinez and Carreño, (1991) refer to its creeping habit, its origin is in southern Canada, Central and Eastern U.S.A., its roots are easy to root and have fine roots, yellow in color and tend to develop superficially, being a great wood producer. Riparia Gloire is the most widely propagated variety of V. riparia. This species resists downy mildew and phylloxera, frost and is very susceptible to calcium carbonate in the soil, does not resist drought, and has a medium resistance to nematodes. Riparia gloire is suitable with European V. vinifera vines, bringing forward fruiting with satisfactory fruit size and quality. It is adapted to porous, well aerated, high humic and humid soils.

2.12.2. VIitis rupestris.

Galet, (1979) considers it to have high resistance to phylloxera, downy mildew, powdery mildew and frost, the shoots root easily and the vines are moderately vigorous when grown in sandy and humid soil, it is more tolerant to calcareous chlorosis but is unsuitable for soils with high pH. It is more drought tolerant than V. Riparia and tends to be less early in both budbreak and fruit ripening.

2.12.3. Vitis berlandieri.

Howell, (1987) mentions that it is native to the Southwest of the U.S.A., in Texas. The resistance to phylloxera is good as well as to diseases and highly resistant to chlorosis. It also resists drought, but has some difficulties in rooting. In general, varietal grafts show good affinity with this rootstock, developing somewhat slowly at first but acquiring good vigor over the years. With this rootstock, fruiting is regular and abundant, achieving an early ripening of the grapes. The effect of this rootstock is that it roots and grafts poorly, but the cross with V. riparia, V. rupestris and V. vinifera produces rootstocks with moderate resistance to phylloxera and tolerance to lime.

III: Conclusion.

The cultivation of the grapevine (*Vitis vinifera*) is considered one of the crops of national and international relevance, where in recent times, production has increased intensely, and the use of suitable varieties in each region where this species can be cultivated. It is fundamental to have good yields and also good quality for the production of red wine, this will depend basically on the selected varieties and the management of the crop from the point of view of the selection of the pattern, selection of the best climate where it is cultivated, also of a good soil and also to give it an adequate management of the plagues and diseases, together with the adequate management of water and nutrition of the plant, this would bring as a consequence, wines of very good quality.

The knowledge and proper management of these production factors will result in excellent wines of a quality, according to international standards, at the end of the road.

IV: Bibliography.

Agustí, M. 2004. Fruticultura. First edition. Editorial Mundiprensa España. pp, 179-197.

Anonymous. 1988. Guía técnica del viticultor. CIAN.SARH-INIFAP-CAELALA. Special Publication N° 25. Matamoros, Coah. 40 p.

Archer, E. y H. C. Strauss, 1985. The effect of plant density on root distribution of three-year-old grafted 99 Richter grapevines, S. Afr. J. Enol Vitic; 6:pp 25-30.

Archer, E. 1988. Effect of plant spacing and trellising systemson grapevine root distribution in J.L. Van Zyl (comp). The grapevine root and its environment, ARC Infruitec-Nietvoorbij, Private Bagh X5026, 7599 Stellenbosch, South Africa, pp. 74-87.

Asociación Nacional de Vitivinicultores A.C. 2008, online http://www.diariodelvino.com/notas3/noticia1257_08feb08.htm (accessed 12/09/12).

Boulay, H. 1965. Arboriculture and Fruit Production. From AEDOS. Barcelona, Spain.401.p.

Bravo, J., 2010. Table grape market. (Online): http://www.odepa.gob.cl/odepaweb/publicaciones/doc/2405.pdf.fecha accessed on: 18/05/2012.

Calderón, E. A., 1977. Fruticultura General. Editorial ECA. pp. 759.

Calderón, E.A. 1998. Fruticultura General.3rd edition. Editorial Limusa. Mexico D.F. 250 p.

Carbonneau, A. y G. Cargnello. 1999. Dictionnaire des systemes de conduit de la vigne. In: Proc. 11th Meeting of the Study Group for Vine Training Systems (GESCO), 6-12 Junio, Sicilia, and Italia.

Castrejón, S.A. 1975. Artificial inoculation of *Phymatotrichum omnivorum* grapevine under greenhouse conditions. CIANE-Laguna, Phytopathology Subproject. Viticulture Research Group. UPM- 2012. Morphology of grapevine. pp.20-30.

Cetto, L. A., 2007. Wines in Mexico. Viticultura. (Online) http://jcbartender.blogspot.mx/2007/08/viticultura-5-los-vinos-en-méxico.html (accessed) 12/02/13.

Chávez, G. W. and P. A. Arata. 2004. Pest and Disease Control in Grapevine. Southern Regional Program Caraveli Operational Unit. Malaga Spain. pp. 18.

De La Cruz and Mario A, Peniche Martinez Ramon A, Roman Becerril A., Enrique A., Ortiz 2015. Physics and Chemistry Of Red Wines Produced In Queretaro, Fototec Magazine, Vol.35, N 5.

Delgado, G. G., 2012. Effect of rootstock vigor on grape production and quality in the variety Shiraz (Vitis vinifera L.), in the region of Parras, Coah. Thesis work UAAAN UL. 60 p.

El siglo de Torreón (2012) http://www.elsiglodetorreon.com.mx /noticia/404480.disminuye-en-fin-de-ano-la-produccion-de-uvas.html: 10/09/2012.
Ferraro, O. R., 1984. Modern Viticulture. Volume II. Editorial Agropecuaria Hemisferio Sur. Montevideo, Uruguay. 180 p.

Galet, P. 1979. Practical Ampelography Grapevine Identification. Cornell University. Press. U.S.A. 60 pág.

Galet, P. 1985. Precis d'Ampelographie Pratique. Imprimerie Ch. Dehan. Montpellier, France. 160 p.

Garritz, A., (2011). "Disclosure: The Chemistry of Wine. 20 pp.

Galet, P. 1988. Grape varieties and vineyards of France. Tome 1. Les Vignes Américaines. ImprimerieCharles   Dehan. Montpellier, France. 85 pág.

Galet, P. 1990. Grape varieties and vineyards of France. Tome II. LÁmpelographie Francaise. II Edition. Imp. Charles Dehan. Montpellier, France. 78 pág.

GIV 2015, Grupo de investigación En Viticultura, UPN, Morphology of grapevine (vitis vinifera), accessed 03 December 2015 online.

González, R. H., 1999. Use of rootstocks in wine grapevines. Informativo La platina. Number 6. Agricultural Research Institute, La Platina Regional Research Center, Ministry of Agriculture. November, Santiago, Chile. 250 p.

González, R.H., H.I. Muñoz, 2000. Use of rootstocks in wine grapevines: general aspects. Agricultural research institute Centro Regional de Investigación La Platina. Ministry of Agriculture. Santiago de Chile.pp.22-23.

Greenspan, M.D. 1994. Developmental changes in the diurnal water Budget of the grape Berry exposed to water deficits, Plant, Cell and Environment.

Hartman, H .T and D .E. Kester.1979. Plant Propagation. Principles and Practices. Compañía editorial Continental S.A. Mexico. 250 p.

Hidalgo, L. 1975. Rootstocks in Viticulture. INIA, booklet number 4. Madrid 11.40 pág.

Hidalgo, L. 2002. Treatise on general viticulture. Third edition, Mundi-Prensa Mexico. 200 p.

Howell, G.S. 1987. Vitis Rootstocks. Chapter 14 in Rootstock for fruticrops. Edited by Romm, R.C., and Carlson, R. F. A. Wilky interscience Publication. pp. 472.

Joachim, G. 2004, Extractive Industries in Arid Zones, Environmental Planning and Management 25 pp.

Lacey, M.J., Allen, M.S. y Harris, R.L.N. 1991. Methoxypyrazines in Sauvigon blac grapes and wines, Am J. Enol Vitic 42:pp103-108.

Larrea, R. A., 1981. Basic viticulture. 1st. Edition. Editorial AEDOS. Spain. pp.82.
Ljubetic, D. 2008. Rootstocks for table grapes: The basis of successful fruit growing. Agricultural Network. [Online]. http://wwwredagrícola.com/view/67/32/. Accessed October 24, 2009.

López-Aguilar, M. 2018. Evaluation of five grape cultivars for red wine in the Comarca Lagunera of Durango. 2nd Evaluation). Bachelor's thesis. UAAAN UL. Torreón Coahuila Mexico, 53 p.

Madero, T. E. 1997. Use of phylloxera resistant rootstocks in vineyards of the Lagunera Region. INIFAP. Deployable for producers No. 2. pp. 2 and 3.

Martínez, C. A., M.A. Elena, Carreño J. E. and Fernández J. R., 1990. Patterns of grapevine.

Morales-Salazar, A. 2020 Vinícola, Cuatro Ángeles. Torreón Coahuila, Mexico. Personal communication.

Martínez, C.A.; Carreño E. 1991. The choice of rootstock in table grape cultivation. Vitivinicultura. Number 11-12. Spain. pp. 59-61.

Martínez, T. F.1991. Biology of the grapevine. Biological foundations of viticulture. Mundi-prensa. 346.

Meraz Ruiz L., 2013, La trascendencia histórica de la zona vitivinícola de baja california, multidisciplinar, n.16.

Morales, P. 1995. Grape cultivation. Technical bulletin No. 6. Second edition. (Online): http://www.zamorano.edu/gamis/frutas/uva.pdf. Accessed on: 09/13/2012.

Mortensen, 1939. Nursery tests with grape rootstock. A. Soc. Hort. Sci. pp. 155 157.

Muñoz, H. I. 2002. Use of rootstocks in grapevines. El Tambo Nursery. Technical information, second part. San Vicente Tagua, Chile. [On line].

Perez, A. Yenia 2015, Vine cultivation current perspectives, accessed December 03, 2015 online.

Reynier, A. 2001. Manual of viticulture. 6th edition. Mundi-Prensa-Mexico. pp. 47, 76-77.

Robles, J. M., J. A. Márquez, R. A. Armenta, and E. Valenzuela. 2004. Diagnosis industrial grapevine chain. Inifap. pp. 28-29.

Rodríguez-Vázquez, Y.A. 2018. Evaluation of five grape cultivars for red wine in the Comarca Lagunera of Durango. (First Evaluation). Bachelor's thesis. UAAAN UL. Torreón Coahuila Mexico, 42 p.

Rodríguez, L. P. 1996. Pests and Diseases of Grapevine in the Canary Islands. Plant Health Section. 3rd edition. pp. 8 and 9.

Ruiz, H. M., 2000. Pests and diseases of grapevine in the Canary Islands. Plant health section. 3ª. Edition. pp. 8 and 9.

Salazar, M.D, and M.P. Melgarejo, 2005. Viticulture, grapevine growing techniques, grape quality and wine attributes. 1st edition. Editorial Mundi-prensa. Madrid. Spain.

Tarango, A.L.A.2015, Problemas y alternativas de desarrollo de las zonas áridas y semiáridas de México, Centro Regional De Estudios De Las Zonas Áridas Y Semiáridas (CREZAS), Accessed December 10, 2015(online).

Ticó, J. and L. 1972. Como ganar dinero con el cultivo de la vid. Ediciones Cedel, Barcelona Spain. 60 p.

Togores, J., H. 2006. The quality of wine from the vineyard. Editorial Mundi-prensa, Mexico, D.F.P. 18, pp 43-46.

Vargas, A.I., V.A. Contreras, M.J. Hernández, and T.A. Martínez, 2006. Arylselenophosphates with selective antifungal action against Phymatotrichum omnivorum. Revista Fitotecnia Mexicana 27. pp. 171-174.

Villanueva, V. N. (2011). "Asociación agrícola de productore de uva de mesa modelo de la agricultura moderna en México en el siglo XXI." Programa de Documentacion de Caso de Exito.

Weaver, R.J. 1976. Grape Growing A. Wiley-Interscience publication New York USA. 220 pag.

Weaver, R. J., 1985. Grape cultivation. Editorial Continental. Mexico .pp. 54, 55, 61, 64.

Winkler, A.J., 1970. Viticulture. 2nd Edition. CECSA. Mexico. 240 p.

Winkler, A.J. 1980. Viticulture. Ediciones CECSA, Davis Ca. USA. 200 pag.

I want morebooks!

Buy your books fast and straightforward online - at one of world's fastest growing online book stores! Environmentally sound due to Print-on-Demand technologies.

Buy your books online at
www.morebooks.shop

Kaufen Sie Ihre Bücher schnell und unkompliziert online – auf einer der am schnellsten wachsenden Buchhandelsplattformen weltweit! Dank Print-On-Demand umwelt- und ressourcenschonend produziert.

Bücher schneller online kaufen
www.morebooks.shop

KS OmniScriptum Publishing
Brivibas gatve 197
LV-1039 Riga, Latvia
Telefax: +371 686 204 55

info@omniscriptum.com
www.omniscriptum.com